SILENT KILLER!

How exposing the eyes

to ordinary light in the hours before bedtime,
or during the night, is doubling the incidence of
breast and prostate cancer.

RICHARD L. HANSLER, PhD

ISBN: 1511984589
ISBN 13: 9781511984584
Library of Congress Control Number: 2015907211
CreateSpace Independent Publishing Platform,
North Charleston, South Carolina

CONTENTS

ACKNOWLEDGMENTS

Thanks go first of all to my wife, Wanda, for her patience and support while I try to spread the word about the benefits of melatonin. Next, I need to thank my partners in this venture: Dr. Ed Carome, Vilnis Kubulins, Dan Carome, and Dr. Marty Alpert. It continues to be great fun trying to make this academic study into a business that helps people.

Many on the faculty and staff at John Carroll University have been extremely helpful in so many ways, including Dr. Joe Trivisonno, Dr. Sally Wertheim, Dr. Joe Miller, Dr. Mary Beadle, Dr. Roy Day, Ms. Diega Bravo, Ms. Carol Clark, and Ms. Tonya Strong-Charles. I also want to thank and acknowledge the support of dear friends Dr. Mike and Mary Michael.

I also want to thank the forty-some students who worked in the Lighting Innovations Institute over the past eighteen years for the inspiration they have provided. I also want to thank my colleagues and friends from my years at GE: John Davenport, Gary Allen, and Roger Buelow. I need to thank and acknowledge the important role that Rev. Dick and Susie Sering have played in my life. Their complete commitment to helping others has been an inspiration to thousands. In Dick's many-year fight against cancer, he inspired me to continue my search for a way to overcome this menace to the health of so many.

I also want to thank my children and grandchildren for their love and support that have helped to create the climate where I have been able to continue working well past normal retirement age. I also want to acknowledge and thank my son-in-law Mark Thomsen for the design of the cover of this book.

INTRODUCTION

My first book, *Great Sleep! Reduced Cancer!* in 2008, has a question on the bottom of its cover: "Are electric lights killing you?" The implied answer was yes. The tone of that book, however, was simply something like "Let's look at the evidence that using light at night is damaging to sleep and to health. The reader can make up his or her mind whether to do something about it." The hope was that others would join the chorus and confront the lighting industry about their unsafe product and at least demand a warning label. That hasn't happened, so I now feel I need to make my point perfectly clear. **People are dying because they have used light bulbs exactly as intended. Light bulbs are silent killers. It's time to let everyone know.** Everyone has the right to know this. I think the federal government should issue a warning, but so far, this hasn't happened. I'm taking on this responsibility myself by writing this book.

How serious is the problem? About 230,000 breast cancers are discovered each year in the United States. About forty thousand women here die annually from the disease. About one in every eight women will be diagnosed with breast cancer during her lifetime. About 220,000 men will be diagnosed with prostate cancer each year in the United States. About twenty-eight thousand men die each year from it. How many will die from heart disease, diabetes, and other types of cancer as a result of exposing the eyes to light at night? Probably double the figure for breast cancer.

So much for the bad news. The good news is that there are light bulbs that don't make people sick and tired (and fat), and that if we use amber eyeglasses, ordinary lights become harmless. I also discuss what modest efforts the lighting industry has made to respond to the problem.

CHAPTER 1

What is the Evidence?

How do we know that using light at night is a silent killer? That's a serious charge. It clearly is not obvious, or I wouldn't have to write this book.

For about the first hundred years of our use of electric light at night, the problem almost escaped detection. This is partly because early light bulbs had carbon filaments and operated at relatively low temperature. This meant they didn't produce much blue light. They were also low wattage and therefore didn't make much total light, and people didn't have very many of them. For more than fifty years, it has been known that exposing the eyes to light suppresses the production of melatonin, the sleep hormone. In 2001, it was discovered that it is primarily the blue rays in ordinary white light that cause the problem (white light is a mixture of all the colors of light).

Each improvement in lighting that made it brighter, bluer, and cheaper increased its impact on health. The gradual change from incandescent to fluorescent and now to LED has worsened the problem. When Henry Ford decided he needed to make cars all night long, he started a trend. Today we have a society in which things go on around the clock.

To be clear, the light itself does not cause harm. The problem is that light stops the body's production of melatonin. A lack of melatonin results in difficulty falling asleep and increases the risk of many deadly diseases.

The most obvious symptom of a lack of melatonin is sleep problems, which are on the rise. More than a third of the people questioned report they have trouble sleeping. Some people simply report they don't have time to sleep. During the work week, they only get a few hours each night and then sleep really late on Saturday and Sunday. Another large group consists of people who do shift work, for whom sleep is a really serious problem. But even people who have ordinary schedules and have the time for sleep still have problems, like having a hard time falling asleep or waking up in the middle of the night and not being able to get back to sleep until it's almost time to get up.

The American Academy of Sleep Medicine accredited its 2,500th sleep center in 2012. This is a fivefold increase in the number of sleep clinics in just ten years. Much of this rise is associated with the need to diagnose and treat people who sleep poorly because of sleep apnea (trouble breathing while asleep). The rise in the incidence of sleep apnea results from the increase in obesity, of which sleep apnea is a side effect. Obesity in turn is at least partly the result of using light at night. The statistics are very clear: the fraction of people who are overweight or obese is significantly higher among night-shift and rotating-shift workers. How light at night and loss of melatonin affects obesity is discussed in my book, "Another Weight Loss Gimmick? Maybe Not!" It involves melatonin's ability to stimulate the formation of brown fat that can convert fat to heat.

Even more deadly diseases than obesity are breast cancer and prostate cancer. It was a study of the incidence of breast cancer among nurses who worked night shift for many years that ignited serious concern among scientists about the dangers inherent in using light at night. I know it alarmed me.

For more than forty years, I did research for GE Lighting on how to produce light more efficiently and on how to manufacture light bulbs of higher quality at lower cost. It wasn't until I retired from that career and began working at John Carroll University that I began to look at the effect of light on health. The studies by Dr. Schernhammer at Harvard Medical School really disturbed me. She is an epidemiologist and had available to her a nurses' health study in which every year, thousands of female nurses fill out an extensive form concerning not only their health but details about their lifestyles. Many also supply blood and urine samples. This information is compiled and made available to research scientists. When Dr. Schernhammer was a student in Germany, she had friends who were nurses who developed breast cancer. She wondered if it had anything to do with their working night shifts.

When she arrived at Harvard, she had a chance to pursue the question. What she found was reported in a 2001 paper entitled "Rotating night shifts and risk of breast cancer in women participating in the nurses' health study." She found that the risk was substantially increased (by 38 percent) for nurses who had worked night shift for many years. She mentioned that it might be related to the loss of melatonin caused by nighttime exposure to light.

I was surprised by this. Did light really have this significant an effect on health? Could exposure to ordinary light make you sick? It didn't seem likely. I began searching the literature in earnest.

Authors note:

When the manuscript for this book was edited the editor suggested that references be cited for studies such as the above. Rather than encumber the book with references it is my choice to simply provide the author's name and initials and date of publication and the reader can quickly access the abstract and frequenty the entire paper by searching www.pubmed.gov.

The next study I read about was from 1991 by R. A. Hahn, entitled "Profound bilateral blindness and the incidence of breast cancer." He found that profoundly blind women were half as likely to have breast cancer as women who were not. One might think that such a dramatic effect would cause breast cancer researchers everywhere to try to figure out what gives blind women this great advantage. However, it wasn't until eighteen years later (2009) that someone did. E. Schernhammer published "Total visual blindness is protective against breast cancer." She found that the risk was slightly less than half as great for totally blind women than for blind women who still experienced control of their circadian rhythms and suppression of melatonin by exposure to light. In this paper, she suggested that increased melatonin and circadian synchronization may play a role in reducing the incidence of breast cancer. I was intrigued by this.

Biological or internal clocks are found in almost all living things, and in many cases, they control when melatonin is produced. Most biological clocks are circadian (about a day) clocks that are kept in synchrony with the sun by exposure to light. In higher animals, the clocks are reset every dawn with such exposure. Twelve hours later, the clocks signal the pineal gland to start producing melatonin. The concentration builds to a maximum about six hours later and then drops to near zero in another six hours. In diurnal animals, melatonin is a signal to receptors throughout the body that it's time to slow down, rest, and rebuild. This beautiful pattern is now being destroyed by electric lighting, the silent killer.

I began looking at studies of how light exposure affects animals and cancer and found dozens of studies dating back many years. A 1994 study using female adult deer mice divided them into two groups. One was housed in long nights (eight hours of light and sixteen hours of darkness), the other in short nights (sixteen hours of light and eight hours of darkness). The short night schedule is the one most humans experience (eight or less hours of darkness and sixteen or more hours of light).

After eight weeks, the mice were all injected with a carcinogen that causes breast cancer. After another eight weeks, they were examined. Of the short-night mice, 90 percent had breast cancer. **None** of the long-night mice had it. What was happening during that long night that could cause such a huge difference? A likely prospect was the time when melatonin was present. Although many rodents are nocturnal (active at night), they produce melatonin during darkness, just like humans. For them, melatonin is the signal to wake up and get going.

A series of papers from the 1980s found that exposure to continuous light had the same effect of increasing the incidence and growth of breast cancer tumors in small rodents as removing the pineal gland that produces melatonin. Adding melatonin to these animals' drinking water prevented the growth of breast-cancer tumors. These experiments were done with rats, mice, and hamsters.

Many animal studies have looked at the effect of light on melatonin production. Dr. George Brainard, who is a leading researcher in this field, began his career back in 1978 with a paper about the pineal gland and melatonin. In 1984, he discovered that blue light was most effective in suppressing melatonin in the Syrian hamster. Seventeen years later (2001), he published results for a human study. It is blue light that is most effective in suppressing melatonin. Published in the same issue of the same journal were the results from a group at the University of Surrey in the United Kingdom. They found identical results. The peak response is at about 470 nm. This is different from the response of the rods and cones. Both groups suggested the control of the circadian system and of melatonin is through sensors in the retina not previously identified. They are a different type of cell from the rods and cones: intrinsically photosensitive retinal ganglion cells (ipRGC). They do not contribute to vision. The nerves from the ipRGC go to the internal clock and to the part of the brain controlling the size of the iris of the eye.

I was very happy when I learned about this. It meant we could do something about the problem of light at night causing illness and death. All we needed to do is to get rid of the blue light at night. Deaths from breast cancer would be cut in half—to the level of totally blind women. Cancer doctors would have to find different jobs. Komen could switch to raising money for research on brain cancer. But life is not like that.

Those who one might think would have the most interest in the remedy appeared to show none.

Another thing I learned was how the incidence of breast cancer had increased over the years in parallel with the use of electric lighting. This did not occur everywhere but only in advanced countries where electricity was available. The incidence in the United States was among the highest of any country, and rich women had a higher incidence than poor women. In less developed countries in Africa and Asia, the incidence of breast cancer is about one-fifth of what it is in the United States.

So far, we have described the evidence that light at night increases the risk of breast cancer that comes from studies of small animals and from epidemiological studies of nurses and blind women and primitive societies. The most convincing studies (to me) are those carried out by a team led by Dr. Blask at Tulane University School of Medicine. It is based on a technique they developed in the early years of this century to grow human breast cancers on the backs of rats but in which the tumor is provided with human blood from volunteers.

In 2005, this team published the first paper describing the results of these novel studies. It showed that when the human blood contained melatonin, the human breast-cancer grafts did not grow or grew slowly, while if the blood did not contain melatonin, the grafts grew rapidly. The human blood with melatonin was obtained from women volunteers during the night when they were in darkness. The blood without melatonin was obtained during the day or during the night after the volunteers had been exposed to bright light.

It was this study, coupled with other evidence, that compelled the International Agency for Research on Cancer (IARC), a part of the World Health Organization (WHO), in 2007 to classify light at night that disrupted the circadian rhythm as a probable carcinogen. When this is mentioned in the press, it usually is regarded as only applying to shift workers. The point that is missed is that anyone using electric lights is disrupting their circadian rhythm and reducing the time when melatonin is present and that the amount produced is reduced compared to what it could be if we didn't have electric lights.

In a 2011 paper, Dr. Blask provided the results of the ongoing study of cancers grafted to rats and fed with human blood. The paper describes various mechanisms by which avoiding light at night to maximize melatonin prevents the growth of estrogen-positive breast cancers. While the process is complex, the result is simple. Avoiding blue light for a few hours before bedtime to maximize melatonin will retard cancer growth.

Commenting on the significance of these findings, Blask says: "The mutual reinforcement of interacting circadian rhythms of melatonin production, the sleep/wake cycle and immune function may indicate a new role for undisturbed, high quality sleep, and perhaps even more importantly, uninterrupted darkness, as a previously unappreciated endogenous (natural) mechanism of cancer prevention."

For the serious reader who wants to examine the studies that establish that using light at night does indeed double the incidence of breast and prostate cancer, the second edition of my book *Great Sleep! Reduced Cancer!* provides an examination of the literature without getting too technical. An alternative is to go to the US government website www. pubmed.gov and search with keywords like *breast cancer, light,* or *melatonin,* and you'll find hundreds of abstracts of technical papers. A Google search will bring up many articles in newspapers and magazines that describe the damage to sleep and health caused by light at night, especially blue light. Most, however, fail to point out the potential decrease in the risk for cancer from maximizing the production of melatonin by avoiding blue light for a few hours before bedtime.

CHAPTER 2

Technology to the Rescue

In 2005, a group at the University of Toronto carried out an experiment in which subjects wore orange goggles while they worked a simulated night shift under bright lights. They measured the amount of melatonin in subjects' saliva every two hours. It turned out that they produced melatonin just as they had during an earlier night when kept in darkness. When they worked a night shift under the same lights without the glasses, they made almost no melatonin. The goggles blocked all of the light at wavelength below about 530 nm (i.e., blue light).

This demonstrated that the problem of melatonin suppression at night could be solved without having to give up all light—just the blue rays.

In that same year, 2005, my group of physicists at John Carroll University succeeded in developing several types of light bulbs that don't produce blue light. These included incandescent, compact fluorescent, and LED lights. To make them available to the public, we formed a spin-off company, Photonic Developments LLC, and opened a website: lowbluelights. com. We also developed amber eyeglasses that block light below about 530 nm, including the ultraviolet. A variety of styles were offered for those who wear prescription glasses (we called them fitover glasses) and for those who don't (we called them non-fitover glasses).

Sale of these products has grown slowly but steadily. When various electronic devices with glowing screens were developed, like smartphones and tablets, we (Daniel Carome, and Vilnis Kubulins) developed filters to block the blue light coming from these screens. Filters for large-screen TV and computers are also available. Table and floor lamps that can go from white light to low-blue light at the flick of a switch are on the drawing board.

We have a number of distributors located in the United States, the United Kingdom, Canada, and Australia. We also have many affiliates and a number of doctors and clinics that routinely recommend our products for their patients, primarily for sleep problems.

After ten years, we still have almost no competitors. One of the authors of the University of Toronto study, Dr. Robert Casper, is a noted gynecologist specializing in helping couples become pregnant. In an unrelated effort, he has developed eyeglasses (called Zircs) that block blue light below about 480 nm. The peak response of the special sensors is at about 470 nm. As far as I can tell, though, these glasses still are not available for retail purchase. They were tested by nurses working rotating shifts with somewhat favorable results: melatonin was preserved and sleep improved. The situation was somewhat unique, however, in that the fluorescent lights in the hospital were unusual. They had almost no emission between 480 nm and 530 nm. This is not the case for most fluorescent lights. For this reason, the Zirc glasses would probably not prevent melatonin loss under more typical lighting conditions.

The silent killer can be technically eliminated with a range of products that allow us to escape from its damaging effects. Practically, however, since only a tiny fraction of the population knows about the problem or these products, the silent killer is still at large. What to do about it?

While avoiding blue light in the hours before bedtime has been demonstrated to increase melatonin, which retards cancer growth, we cannot claim that avoiding blue light in the hours before bedtime will definitely result in less cancer. To do that would require massive clinical trials, of which the first has yet to be started. Unless the federal government funds such studies, they are not likely to be done. There are no expensive pills that would ordinarily provide the incentive for a drug company to fund a study.

Fortunately, the reader is not constrained by what the government allows us to claim. Because of this limitation, we have largely restricted our claims to the improvement in sleep obtained by restricting blue light. Even our attempts to let people know how our glasses can help them sleep better has received some opposition from the companies that make sleeping pills. We tried to place ads on Google (you pay each time a potential customer clicks on them). Whoever pays the most gets top position on the page. We started with a penny a click, but the drug companies were quickly willing to boost what they pay to five dollars a click. We could not compete.

There is a chance that clinical trials to prevent breast and prostate cancer will be done. The funding might come from drug companies. They would not be based on melatonin, which is not a controlled substance and is regarded as a food supplement by the United States government. Rather, they would be based on one or more of the patented melatonin agonists, molecules chemically similar to melatonin that act on the body much like melatonin. Four have been developed by four different companies. Tasimelteon was developed by Vanda Pharmaceuticals and is being marketed to blind people to help synchronize their circadian rhythm to the daily schedule. Totally blind are "free running" in which their internal clock is producing days longer than 24 hours, like everyone else, but can't be reset by exposure to light every morning. This is called N24 (non 24). Their circadian rhythm gradually progresses around the clock so they will have periods when melatonin is produced during the day making them sleepy, and not at night when they are trying to sleep. It was demonstrated back in the 1990s that giving a small dose of melatonin at the same time every day will synchronize the circadian rhythm. Doctors cannot prescribe melatonin but they can prescribe Tasimelteon and no doubt insurance companies will pay for it. It seems likely that the companies will try to test their drugs to try to provide other of the beneficial effects of melatonin, including its cancer fighting ability.

CHAPTER 3

Cancer as a Balancing Act

Why doesn't everyone get cancer? Why does the incidence increase with age? What causes cancer? What can prevent cancer? Is there a tipping point? These are important questions, and learning the answers can help the reader to make better decisions. We will look at these questions in detail. (The answers below are the opinions of the author.)

Why doesn't everyone get cancer?

Some scientists believe we all have cancer but that it doesn't always grow large enough to be detected. Some of the evidence comes from statistical studies of the benefit of mammographic screening. Since about 2000, there has been an ongoing debate whether screening should begin at age fifty or age forty. Comparing deaths from breast cancer in large populations having the two different starting ages for screening, found almost no decrease in deaths from breast cancer when the screening started at the younger age. One explanation was that many of the tumors found in the early screening were so small and grew so slowly that they never would have became lethal.

Since there are so many causes for cancer from which no one can escape (e.g., cosmic rays), it seems likely we all have dormant cancers. Studies of metastatic cancers that become active after many years of dormancy provide further support for this theory.

Why does the incidence of breast cancer increase with age?

Every time a cell divides, there is a chance that errors will occur as the genes are replicated. Gradually, as errors accumulate, the chance that they will result in a cell becoming malignant increases. The chance of errors increases if reactive oxygen species (ROS) are present. ROS are caused by various kinds of stress including radiation and toxins. Melatonin has the ability to eliminate ROS. At the same time that the creation of micro cancers increases with age, the amount of melatonin produced each night decreases with age, according to a number of studies. The combination of these factors explains the increasing incidence of cancer with age. Clearly, anything that increases melatonin supply can help to shift the balance in your favor.

What causes cancer?
Radiation

Radiation is a significant source of cancer initiation. In the nuclear age, we are subject to increasing amounts of man-made radiation. Many of the radioactive molecules injected into the earth's atmosphere during the testing and use of atomic and hydrogen bombs during WWII had long half-lives and are still around. The recent tsunami that wiped out the reactor in Japan created deadly levels of radiation in parts of Japan. Nuclear power plants are always leaking some radioactive species into the atmosphere.

Naturally produced radon gas coming from deep in the ground is common in many parts of the world and can accumulate to dangerous levels in well-insulated homes. The ordinary chest X-ray does not produce much radiation damage. However, an MRI or CAT scan subjects the patient to much higher levels of radiation. Cosmic radiation and radiation from solar flares is doing its share of damage to our genes, especially during high-altitude flights. Because most of this radiation is unavoidable, it is more important to maximize the things that we can do to prevent cancer from growing—for example, the production of melatonin.

Toxic Substances

The IARC WHO list of the toxic substances that have been shown to cause cancer is a long one. Class I is known carcinogens:

- Acetaldehyde

- 4-Aminobiphenyl

- Aristolochic acids, and plants containing them

- Arsenic and arsenic compounds

- Asbestos

- Azathioprine

- Benzene

- Benzidine

- Benzo[a]pyrene

- Beryllium and beryllium compounds

- Chlornapazine (N, N-Bis(2-chloroethyl)-2-naphthylamine)

- Bis(chloromethyl)ether

- Chloromethyl methyl ether

- 1,3-Butadiene

- 1,4-Butanediol dimethanesulfonate (Busulphan, Myleran)

- Cadmium and cadmium compounds

- Chlorambucil

- Methyl-CCNU (1-(2-Chloroethyl)-3-(4-methylcyclohexyl)-1-nitrosourea; Semustine)

- Chromium(VI) compounds

- Ciclosporin

- Contraceptives, hormonal, combined forms (those containing both estrogen and a progestogen)

- Contraceptives, oral, sequential forms of hormonal contraception (a period of estrogen-only followed by a period of both estrogen and a progestogen)

- Cyclophosphamide

- Diethylstilboestrol

- Dyes metabolized to benzidine

- Epstein-Barr virus

- Estrogens, nonsteroidal

- Estrogens, steroidal

- Estrogen therapy, postmenopausal

- Ethanol in alcoholic beverages

- Erionite

- Ethylene oxide

- Etoposide alone and in combination with cisplatin and bleomycin

- Formaldehyde

- Gallium arsenide

- *Helicobacter pylori* (infection with)

- Hepatitis B virus (chronic infection with)

- Hepatitis C virus (chronic infection with)

- Herbal remedies containing plant species of the genus *Aristolochia*

- Human immunodeficiency virus type 1 (infection with)

- Human papillomavirus type 16, 18, 31, 33, 35, 39, 45, 51, 52, 56, 58, 59 and 66

- Human T-cell lymphotropic virus type 1

- Melphalan

- Methoxsalen (8-Methoxypsoralen) plus ultraviolet A radiation

- 4,4-methylene-bis(2-chloroaniline) (MOCA)

- MOPP and other combined chemotherapy including alkylating agents

- Mustard gas (Sulfur mustard)

- 2-Naphthylamine

- Neutron radiation

- Nickel compounds[2]

- 4-(N-Nitrosomethylamino)-1-(3-pyridyl)-1-butanone (NNK)

- N-Nitrosonornicotine (NNN)

- Opisthorchis viverrini (infection with)

- Outdoor air pollution

- Particulate matter in outdoor air pollution

- Phosphorus-32, as phosphate

- Plutonium-239 and its decay products (may contain plutonium-240 and other isotopes), as aerosols

- Radioiodines, short-lived isotopes, including iodine-131, from atomic reactor accidents and nuclear weapons detonation (exposure during childhood)

- Radionuclides, α-particle-emitting, internally deposited[5]

- Radionuclides, β-particle-emitting, internally deposited[5]

- Radium-224 and its decay products

- Radium-226 and its decay products

- Radium-228 and its decay products

- Radon-222 and its decay products

- *Schistosoma haematobium* (infection with)

- Silica, crystalline (inhaled in the form of quartz or cristo-balite from occupational sources)

- Solar radiation

- Talc containing asbestiform fibres

- Tamoxifen

- 2,3,7,8-Tetrachlorodibenzo-para-dioxin

- Thiotepa (1,1,1-Phosphinothioylidynetrisaziridine)

- Thorium-232 and its decay products, administered intrave-nously as a colloidal dispersion of thorium-232 dioxide

- Treosulfan

- *ortho*-Toluidine

- Vinyl chloride

- Ultraviolet Radiation

- X-Radiation and Gamma radiation

Mixtures:

- Aflatoxins (naturally occurring mixtures of)

- Alcoholic beverages

- Areca nut

- Betel quid with tobacco

- Betel quid without tobacco

- Coal-tar pitches

- Coal-tars

- Household combustion of coal, indoor emissions from

- Diesel exhaust

- Mineral oils, untreated and mildly treated

- Phenacetin, analgesic mixtures containing

- Plants containing aristolochic acid

- Polychlorinated biphenyls

- Salted fish (Chinese-style)

- Shale-oils

- Soots

- Tobacco products, smokeless

- Wood dust

The IARC list for Class 2a (probable carcinogens) is even longer, and I will not provide it here. It ends with a list of **exposure circumstances**:

- Art glass, glass containers and pressed ware (manufacture of)

- Carbon electrode manufacture

- Cobalt metal with tungsten carbide

- Hairdresser or barber (occupational exposure as a)

- Petroleum refining (occupational exposures in)

- Shift work that involves circadian disruption

Note the last item in this list. This is the only semigovernmental warning to the general public of the existence of the silent killer. After more than twenty-five years of scientists issuing warnings, one might hope for a little more.

A number of scientists believe that the disruption of the circadian rhythm is the reason that working night shift increases the incidence of breast and other cancers, not the loss of melatonin. However, a clear mechanism for this has not been established. During a normal circadian cycle, melatonin travels throughout the body during darkness, resulting in many changes in the behavior of the tissues and organs. Body temperature drops, many processes slow down, and damages are repaired. When the normal cycle is interrupted, local stress occurs, resulting in an increase in ROS that could increase risk of new cancers developing. The lack of melatonin may be the most damaging effect of the disruption.

What can prevent cancer?

The most powerful way to prevent cancer is to avoid producing a cell with a defective genome—one that has experienced an error in the copying of the DNA during cell division. Many of the items in the above list of causes do so by damaging the DNA of cells either directly or by creating reactive oxygen species (ROS) that attack and damage the

DNA. We know now that melatonin is one natural material that destroys ROS. The body produces many other antioxidants that eliminate ROS. The reason healthy eating is so important is that many foods contain antioxidants. The foods in the following list fight cancer by providing antioxidants that eliminate ROS.

Rank	Food item	Serving size	Total antioxidant capacity per serving size
1	Small red bean (dried)	Half cup	13,727
2	Wild blueberry	1 cup	13,427
3	Red kidney bean (dried)	Half cup	13,259
4	Pinto bean	Half cup	11,864
5	Blueberry (cultivated)	1 cup	9,019
6	Cranberry	1 cup (whole)	8,983
7	Artichoke (cooked)	1 cup (hearts)	7,904
8	Blackberry	1 cup	7,701
9	Prune	Half cup	7,291
10	Raspberry	1 cup	6,058
11	Strawberry	1 cup	5,938
12	Red Delicious apple	1 whole	5,900
13	Granny Smith apple	1 whole	5,381
14	Pecan	1 ounce	5,095
15	Sweet cherry	1 cup	4,873
16	Black plum	1 whole	4,844
17	Russet potato (cooked)	1 whole	4,649
18	Black bean (dried)	Half cup	4,181
19	Plum	1 whole	4,118
20	Gala apple	1 whole	3,903

(WebMD Public Information from the United States Department of Agriculture)

While I encourage everyone to eat healthy foods, I do so with the realization that the evidence of their benefit in the case of cancer is limited.

While they show benefit for heart disease and other diseases, in the case of cancer, the only results that show up clearly in well-designed studies are the negative effects of alcohol, smoking, obesity, and eating red meat. What you put in your body beyond that does not seem to make much difference.

Melatonin is the only antioxidant whose lack has been demonstrated to increase the risk of some cancers in both animal and human studies. Melatonin fights cancer in at least three additional ways. For tumors to grow, they need to be supplied with new blood vessels. Melatonin has been shown to prevent their growth. Melatonin blocks a tumor's metabolism of linoleic acid into a carcinogen (HODE). Melatonin prevents the cancer from locally producing estrogen, which stimulates its own growth.

While I have devoted only one short paragraph to these three additional ways melatonin fights cancer, each may be a more significant factor than melatonins antioxidant capabiities. My book, *Heroes of CancerPrevention Research,* describes how some of melatonin's cancer fighting abillities were discovered.

Is there a tipping point?

It seems logical to me that if we have dormant cancers in our bodies that might start growing and things that impede their growth, a balance does exist. If either one or the other changes, the probability of a detectable cancer emerging will change. It also seems logical that one should do whatever is reasonable to try to shift the balance in one's favor—avoiding things known to cause cancer (pesticides, radiation, smoking, alcohol, obesity, red meat) and maximizing healthy foods and hours of melatonin production (i.e., avoid blue light in the hours before bedtime).

CHAPTER 4

The Bright Side of Light

While I have been blasting the use of ordinary light at night as the silent killer, I recognize that exposing the eyes to light rich in blue rays in the morning is equally important for both good sleep and good health.

Much of the beauty of light lies in its ability to be broken up into its different colors. The human eye is able to distinguish hundreds of different shades. What can be more beautiful than the deep blue of the sky on a perfectly clear day or the rich red of a rose or the clear yellow of a tulip? The whole world of art depends entirely on the ability of the eye to behold it.

Because the eye has been the subject of intense study for centuries, it is truly remarkable that it is only very recently that its other vital function (which is essentially unrelated to vision) should be discovered, namely, the control of the internal master clock and the circadian rhythm. While the basic fact that it is mostly blue light that suppresses melatonin has been known since 2001, the actual complexity of the process is just now being sorted out. The three different kinds of cones that detect red, green, and blue light and the rods that come into play in dim light all play minor roles in the control of the circadian system and melatonin production.

Earlier, it was thought that fairly bright light was required to suppress melatonin. It is now recognized that the light levels found in a typical home are sufficient to significantly reduce melatonin production. However, a powerful benefit of blue light is its ability to increase alertness and performance. Exposing the eyes to light in the morning not only resets the circadian clock but also suppresses any production of melatonin if the cycle has not finished completely. It is also believed to promote the production of cortisol, a stimulant to the nervous system. Some studies have shown that exposing the eyes to lots of light during the day has the effect of increasing the amount of melatonin produced during the night.

Many people need coffee first thing in the day just to get going. If one has a robust circadian rhythm, starting with resetting the clock in the morning and where further resetting is prevented by protecting the eyes from blue light in the evening, one may no longer find a need for coffee. With melatonin gone and cortisol building, no additional boost is required.

The benefit of understanding how the circadian clock works is demonstrated in a press release I issued in 2012, which I reproduce below:

Help Your Teenager Switch from Summer Schedule to School Schedule

Special blue-blocking glasses painlessly trick the body into advancing the circadian cycle to make waking up for school a lot easier. Glasses are available from Photonic Development at http://www.lowbluelights.com.

University Heights, OH (PRWEB) August 20, 2012

During the course of the summer, many teenagers develop the habit of staying up later and later. By now they are staying up 'till 3 or 4 a.m. and sleeping until noon. During this time their internal or circadian clock has adjusted so it is set about five hours later than the clock on the wall. Between now and when school starts it needs to be moved back those 5 hours.

The classic way to adjust the internal clock (1) is to expose the eyes to bright light on waking. It is most effective to move the clock a little at a time. If school starts the day after Labor Day, divide the 5 hours by how many days remain between now and Labor Day to get how much you need to accomplish each day. To use the classic method requires you wake up your youngster that many minutes earlier each morning and get him/her into bright light. Realistically, this just isn't going to happen.

The modern way (2) to advance the circadian clock works on the time when melatonin starts flowing after going into darkness. By moving the time of going into darkness a little earlier each night, the same effect can be obtained. Getting your teen to go to bed earlier each night is also an impossible task.

Fortunately, in 2001, scientists in this country (3) and in the United Kingdom (4) reported an astounding finding. There are special light sensitive cells in the retina of the eye that are different from the rods and cones. They have nothing to do with vision. They control the circadian clock and the production of melatonin. The special cells respond most strongly to blue light. Dr. Kayumov and his colleagues at the University of Toronto found (5) that subjects wearing blue-blocking glasses produced melatonin during the night even though exposed to bright light, just as they had when in darkness on an earlier night.

In 2005, physicists at John Carroll University developed light bulbs that don't make blue light and eyeglasses that block it. They are available through a spin-off company at http://www.lowbluelights.com.

All your student needs to do is put on a pair of the glasses or switch on special light bulbs some number of minutes earlier each night starting a bit ahead of the most recent bedtime.

This will gradually advance the circadian clock to an earlier hour. Whether he/she actually starts going to bed earlier is not required. Because the flow of melatonin is starting a bit earlier each night, going to bed earlier may just happen, without a fight.

Getting up earlier and exposing the eyes to light should be encouraged. You need all the help you can get. Providing breakfast outside is a good strategy for make this happen in a pleasant way.

Few plans work perfectly. The first day of school is still not likely to be easy. The good news is that this trick of using the light bulbs or glasses can be used to keep advancing the circadian cycle to an earlier hour. By getting to where you are putting on the glasses two or more hours ahead of bedtime, you will have lots of melatonin present, by bedtime. This allows sleep to come quickly. Melatonin can only flow for about 10 or 11 hours (6). Starting earlier means ending earlier. Your student may actually be wide awake and ready to learn by the time his/her first class begins.

You may find this trick is also good for you and the rest of the family. There is a very significant side effect. Maximizing melatonin is likely to reduce your risk for breast cancer and, very likely, other cancers (7). There does not appear to be any down side to eliminating blue light in the evening. For millions of years before we had electric lights we didn't have blue light at night. Using special light bulbs or wearing the glasses allows us to go back to that peaceful time.

(1)
Am J Physiol. 1992 Aug;263(2 Pt 2):R428-36. "High-intensity light for circadian adaptation to a 12-h shift of the sleep schedule." Eastman CI.

(2)
Sleep. 2005 Jan;28(1):33–44.

"Advancing circadian rhythms before eastward flight: a strategy to prevent or reduce jet lag."

Eastman CI, Gazda CJ, Burgess HJ, Crowley SJ, Fogg LF.

(3)
J Neurosci. 2001 Aug 15;21(16):6405–12.

"Action spectrum for melatonin regulation in humans: evidence for a novel circadian photoreceptor."

Brainard GC, Hanifin JP, Greeson JM, Byrne B, Glickman G, Gerner E, Rollag MD.

(4)
J Physiol. 2001 Aug 15;535(Pt 1):261–7.

"An action spectrum for melatonin suppression: evidence for a novel non-rod, non-cone photoreceptor system in humans."

Thapan K, Arendt J, Skene DJ.

(5)
J Clin Endocrinol Metab. 2005 May;90(5):2755–61. Epub 2005 Feb 15.

"Blocking low-wavelength light prevents nocturnal melatonin suppression with no adverse effect on performance during simulated shift work."

Kayumov L, Casper RF, Hawa RJ, Perelman B, Chung SA, Sokalsky S, Shapiro CM.

(6)
PLoS One. 2008 Aug 26;3(8):e3055.

"Individual differences in the amount and timing of salivary melatonin secretion."

Burgess HJ, Fogg LF.

(7)
Cancer Lett. 2009 Aug 18;281(1):1–7. Epub 2008 Dec 12.

"Circulating melatonin and the risk of breast and endometrial
cancer in women."

Viswanathan AN, Schernhammer ES.

CHAPTER 5

The Future of Lighting

During the Spring of 2014, the Illuminating Engineering Society sponsored a meeting in Cleveland, for which we submitted a poster. We were allowed to submit a paper describing the poster. I reproduce this paper below.

Lighting for Health: The Dawn of a New Epoch
Edward Carome, Richard Hansler, and Vilnis Kubulins
Lighting Innovations Institute, John Carroll University

Abstract

There is increasing evidence reported in detail in the medical literature that low levels of melatonin, the so-called "sleep hormone", increases the incidence of breast and prostate cancer, diabetes, obesity and other serious medical conditions. It has long been known that the time of production of melatonin by the pineal gland can be as long as 11 to 12 hours, i.e., the time spent in darkness by humans as they evolved, but that production stops when the eyes are exposed to light. Thus, since the advent of artificial lighting in developed countries, melatonin production time has been reduced for most people to the time they spend asleep, i.e., to from 6 to 8 hours. When medical researchers proved in 2001 it is mainly blue light at wavelengths below about 520nm that suppresses melatonin production, it opened a way for lighting developers to possibly improve peoples' health. On the other hand, the new LED white light sources tend to be especially strong in the blue range. LED lighting is being promoted for use throughout the home because of its potentially higher lumen per watt efficacy and much longer life than compact fluorescent lamps and the more widely used incandescent lamps, the production of many of which has been ruled out by the Energy Department. LEDs also form the light sources in the screens of computers and televisions and in the ubiquitous i-Pads and i-Phones and similar hand-held electronic devices that are even taken to bed. Not only has the steadily increasing use of artificial lighting in the evening during the past century possibly contributed to increasing a number of serious medical maladies, the advent of the use of the blue light rich LED type lighting may further acerbate this. In this presentation we consider (a) recent medical research on the health effects of melatonin, (b) how blue light suppresses melatonin production, (c) the intensity levels of blue light in frequently used light sources, and (d) a number available simple ways to modify lighting to avoid melatonin suppression and possibly improve health.

Introduction

Until now the goal of lighting had been two-fold: to enhance human performance and to be aesthetically pleasing. Beginning in the 1990's, scientists have been warning that using ordinary light bulbs at night was increasing the incidence of various very serious illnesses, including breast and prostate cancer. Until 2001 the feeling prevailed that there was not much that could be done about the ill effects, since people were not going to give up using light at night.

In 2001, however, two independent research groups [Brainard 2001, Thapan 2001] reported studies revealing that it is primarily blue light that suppresses melatonin and that there are sensors in the retina (different from the rods and cones) that control the internal clock and the pineal gland that produces melatonin, the so-called "sleep hormone". These newly identified sensors do not contribute to vision (other than partial control of the iris of the eye) and the nerve fibers from them go to the hypothalamus, not the visual cortex.

In 2005 a research group at the University of Toronto [Kayumov 2005] did an experiment in which subjects worked a simulated night shift under bright lights while wearing goggles that blocked light at wavelengths shorter than about 530 nm, i.e., the blue part of the visible spectrum. They found that the subjects produced melatonin very much like they had on an earlier night when they were kept in darkness.

Also in 2005, our group at John Carroll University (JCU) developed light bulbs that do not produce blue light and eyeglasses that block blue light [Hansler 2005]. These products were made available at www.lowbluelights.com operated by a spin-off company Photonic Developments LLC. These products have helped improve the sleep of 90% of those who have used them. A double blind sleep study based on their use conducted at JCU showed a significant improvement in sleep quality and in mood [Burkhart 2009].

Melatonin Suppression by Home Lighting

Table I. Calculated Percentage of Melatonin Suppressing Light of Various Sources

Source Type	Percent Melatonin Suppressing Light	Lumen Output
Ecosmart 14W CFL 5000K, 60W Equiv.	41%	922

GE Soft White 43W Halogen 2700K, 60W Equiv.	31%	812
GE Soft White 60W Incandescent, 2800K	29%	840
LS Good Night 12W LED 2500K, 60W Equiv.	22%	918
LowBlueLights 7W LED 1500 K	4%	371

In Table I the percent melatonin suppressing light was computed by multiplying the measured output energy versus wavelength curve by a melanopsin absorption versus wavelength curve, i.e., the best fit to the data presented in [Brainard 2001] and [Thapan 2001].

Recent studies [Gooley 2011, Santhi 2012] have determined that typical home artificial lighting levels in the evening are sufficient to produce a significant drop in melatonin production. With the phasing-out of the incandescent lamp, the question arises concerning the newer light sources and their ability to suppress melatonin. Since it is the blue wavelengths that are most effective in suppressing melatonin, measurements have been made of the fraction of the light that is in the blue weighted by the melatonin suppression curve. Table I shows the results of measurements of the light output of several types of light bulbs, including those claiming "low" amounts of blue light. That this claim is somewhat overly optimistic is emphasized in Figure I, where the output spectra of the latter three type lamps listed Table I are plotted, assuming equal incident energy densities. Note that the Lighting Sciences "Good Night" LED lamp emits close to the blue light of a 60 watt incandescent lamp, i.e., a "typical home artificial lighting" source.

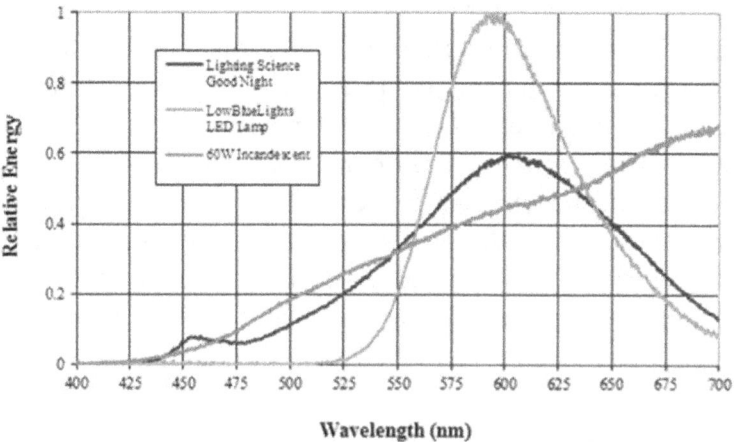

Figure 1. Spectra of three lamps in regions of equal incident energy densities.

The Evidence that Light at Night Damages Health

Animal Studies: The health effects of light were first observed in small rodents that could be raised in an easily controlled light environment. For example, a study of mice [Nelson 1994] was reported in which female mice were divided into one group raised in 16 hours of darkness and 8 hours of light and a second group raised in the reverse schedule, 16 hours of light and 8 hours of darkness. At eight weeks of age they were injected with a carcinogen. After eight additional weeks it was found that about 90% of the long day mice developed squamous cell carcinoma while none of the short day mice developed tumors.

An even earlier study [Kothari 1982] in female rats found 95% of the rats developed mammary tumors when given a carcinogen and raised in continuous light compared to 60% that developed tumors when raised in ten hours of light and fourteen hours of darkness. A 1983 study in hamsters [Stanberry 1983] found that eighteen hour nights increased the time for tumors to start and decreased the rate of growth. They concluded that the quantity, time and duration of melatonin presentation all had an important effect on tumor growth. Many other studies found a similar relationship between hours in darkness, amount of melatonin produced and decrease in the incidence and growth rate of tumors.

Epidemiological Studies: A study of nurses [Schernhammer 2001] who had worked night shift for many years found they had significantly increased risk for developing breast cancer than nurses who had not worked shifts. A related study [Schernhammer 2009] sampled melatonin in overnight urine in women nurses. They were divided into four groups according to the amount of melatonin measured. Eight years later the incidence of breast cancer was determined. Those with the highest amount of melatonin had about half the incidence of breast cancer than those in the lowest melatonin group. In a related study the nurses with the highest amount of melatonin in overnight urine also had a lower incidence of colon cancer.

A study [Hahn 1991] found that totally blind women had about half the incidence of breast cancer as women with normal vision. In a related study [Flynn-Evans 2009] it was noted there are two types of blind women, those who were totally blind and those who had no vision but whose circadian rhythm (and melatonin production) was controlled by light. Breast cancer incidence was found to be half as great in the totally blind. In totally blind women the flow of melatonin is not diminished by exposure to light.

The incidence of breast cancer is much lower (as much as five times lower) in primitive societies that do not have electric lighting than in the modern western societies [Stevens 2002].

Human Breast Cancer Studies: Human breast cancers grown as grafts on animals but provided with human blood, were found [Blask 2005] to grow rapidly if the blood did not contain melatonin, and slowly or not at all, when the blood contained melatonin. The blood with melatonin was obtained from volunteers during the night and the blood without melatonin was obtained from volunteers during the day or during the night following two hour exposure to bright light.

Human Prostate Cancer Studies: A study [Kubo 2006] of men working night shift show an increase in the incidence of prostate cancer for night shift workers compared to day shift workers. Men working rotating shifts had the highest incidence. Human prostate cancers grown as grafts on rats but supplied with human blood grew rapidly when the blood lacked melatonin and slowly when the blood contained melatonin.

Metastasis of Cancer: Metastasis of cancer to distant sites is the thing that kills most cancer patients, not the original tumor. A breakthrough study [Mao 2012] at Tulane and Thomas Jefferson Medical Schools showed that the increased risk of metastasis of both breast cancer and prostate cancer can result from disrupting the circadian (daily) cycle causing loss of melatonin due to exposure to light at night.

The study examined the molecular processes involved in the transition of stable cancer cells into cancer cells capable of moving through the blood stream to distant sites where new tumors can develop. They examined how the presence of various compounds required for the different steps in the process were associated with the presence or absence of melatonin. They looked at this in both cultured cancer cells (both breast cancer and prostate cancer) and also in human cancers grown as grafts on the backs of rats but supplied with human blood. In every case, cancer cells retained a static structure when melatonin was present; however, in the absence of melatonin, the conditions necessary for metastasis to occur were observed.

Seasonal Affective Disorder: The standard treatment for seasonal affective disorder (SAD), or the winter blues, is to expose the eyes to bright light for about a half an hour first thing in the morning. This will cause the pineal gland to stop making melatonin. It has also been demonstrated to advance the start of the flow of melatonin to an earlier hour in the evening. By starting earlier it finishes its flow earlier, since the pineal gland, even in the blind, can produce melatonin for only 11 to 12 hours, i.e., the time spent in darkness as humans evolved. This solves the problem of too much melatonin in the morning. However, exposing the eyes to light in the evening prevents the flow of melatonin from starting and delays the circadian rhythm, essentially undoing what light in the morning has done. By wearing glasses that block blue light allows the flow to start. The average time for melatonin to flow (if the person is in darkness) is 11.4 hours according to a recent study [Burgess 2008]. Putting on glasses at 7 P.M. should allow the flow to be over by 7 A.M. This provides a lasting solution to SAD.

Alzheimer's Disease: In recent papers from Italy [Polimeni 2014], Russia and the United States, scientists have described the possible benefits of melatonin in avoiding and treating Alzheimer's disease. Because of exposure to light in the hours before bedtime, most people only make melatonin for 7 or 8 hours a night. A recent study [Lin 2013] of the spinal fluid that bathes the brain suggests this reduction from the maximum of 11 to 12 hours in the time that the pineal gland makes melatonin (and other antioxidants) may increase the probability for the formation of the plaques associated with Alzheimer's disease. Studies in animals and humans [Leston 2009] show that the concentration of melatonin in the spinal fluid is significantly higher than in the blood. Melatonin is thought to be the unique antioxidant that protects the brain from damage by eliminating the free radicals that can damage the brain cells.

Type 2 Diabetes: Nurses with the highest melatonin production have about half the type 2 diabetes of those with the lowest melatonin, according to a Harvard study [McMullan 2013]. Quoting from the abstract, "Among participants without diabetes who provided urine and blood samples at baseline in 2000, we identified 370 women who developed type 2 diabetes from 2000–2012 and matched 370 controls using risk-set sampling. Associations between melatonin secretion at baseline and incidence of type 2 diabetes were evaluated with multivariable conditional logistic regression controlling for demographic characteristics, lifestyle habits, measures of sleep quality…" Comparing the results, they found that "Women in the highest category of melatonin secretion had an estimated diabetes incidence rate of 4.27 cases/1000 person-years compared with 9.27 cases/1000 person-years in the lowest category." That is, those with the highest melatonin were less than half as likely to develop Type 2 diabetes as those in the lowest melatonin category.

Health Benefits of Light

While the psychological benefits of light are recognized by everyone, the actual physiological effect of light has really only been recognized since 2001 when the special sensors in the eye that control the circadian rhythm were identified. The great importance of early morning exposure of the eyes to light in order to reset the circadian clock is still relatively unknown. Providing light rich in the blue component will be a good starting point for the lighting industry. Studies at Rensselaer, GE Lighting, Philips, Lighting Science and Photonic Developments are beginning to provide specially designed lamps with extra blue light for this purpose. Studies in nursing homes [Figueiro 2013] are showing that providing higher levels of light during the daytime results in better sleep at night.

The ideal light source of the future will be one that can change its spectral content with the time of day. During the day it will be a white source, rich in blue wavelengths and at night will be devoid of blue light but providing all the other colors. Because of the extremely long lifetime of LEDs, it may be possible for some applications to do away with the idea of an easily replaceable bulb in favor of different designs. The switch from day to night lighting may be as simple as a mechanically operated switch, by a remote control, by an internal clock, or a signal from the provider of the electricity.

Having light sources to use at night that do not disrupt the circadian rhythm is especially important for pregnant women and those with new babies [Hansler 2013]. Sleep is difficult enough to get when there is a baby in the house without this unnecessary disruption from exposure to ordinary lighting during the night.

On the other hand, during the day and under special conditions, recent studies [Sahin 2013] have demonstrated that both daytime and nighttime exposure to blue light is effective in improving human performance on tasks requiring alertness. Instead of turning to caffeine for a boost, exposure to light rich in the blue wavelengths provides a drug-free alternative.

Conclusion

The new epoch in lighting in which the health effects of light become the number one concern is not here yet. But, in step with the introduction of LED lighting, it will very likely be making rapid progress in the very near future.

References

Blask DE[1], Brainard GC, Dauchy RT, Hanifin JP, Davidson LK, Krause JA, Sauer LA, Rivera-Bermudez MA, Dubocovich ML, Jasser SA, Lynch DT, Rollag MD, Zalatan F, "Melatonin-depleted blood from premenopausal women exposed to light at night stimulates growth of human breast cancer xenografts in nude rats", *Cancer Res* **65**: 11174–84 (2005)

Burgess HJ, Fogg LF, "Individual differences in the amount and timing of salivary melatonin secretion", *PloS One* **3**:8 (2008)

Burkhart K, Phelps JR, "Amber lenses to block blue light and improve sleep: a randomized trial", *Chronobiol Int* **26**:1602–12 (2009)

Brainard GC[1], Hanifin JP, Greeson JM, Byrne B, Glickman G, Gerner E, Rollag MD, "Action spectrum for melatonin regulation in humans: evidence for a novel circadian photoreceptor", *J Neurosci* **15**: 6405–12 (2001)

Figueiro MG, Lesniak NZ, Rea MS, "Implications of controlled short-wavelength light exposure for sleep in older adults", *BMC Res Notes* **4**: 334 (2011)

Flynn-Evans EE, Stevens RG, Tabandeh H, Schernhammer ES, Lockley SW, "Total visual blindness is protective against breast cancer", *Cancer Causes Control* **20**: 1753–6 (2009)

Gooley JJ, Chamberlain K, Smith KA, Khalsa SB, Rajaratnam SM, Van Reen E, Zeitzer JM, Czeisler CA, Lockley SW, "Exposure to room light before bedtime suppresses melatonin onset and shortens melatonin duration in humans", *J Clin Endocrinol Metab* **96**: 2010–2098 (2010)

Hahn RA, "Profound bilateral blindness and the incidence of breast cancer", *Epidemiology* **2**: 208–10 (1991)

Hansler RL, *Pregnant? New Baby? Need Sleep* (Amazon) (2014)

Hansler R, Carome E. Kubulins V, www.lowbluelights.com

Kayumov L, Casper RF, Hawa RJ, Perelman B, Chung SA, Sokalsky S, Shapiro CM, "Blocking low-wavelength light prevents nocturnal melatonin suppression with no adverse effect on performance during simulated shift work", *J Clin Endocrinol Metab.* **90**:2755–61 (2005)

Kubo T, Ozasa K, Mikami K, Wakai K, Fujino Y, Watanabe Y, Miki T, Nakao M, Hayashi K, Suzuki K, Mori M, Washio M, Sakauchi F, Ito Y, Yoshimura T, Tamakoshi A, "Prospective cohort study of the risk of prostate cancer among rotating-shift workers: findings from the Japan collaborative cohort study" *Am J Epidemiol* **164**: 549–55 (2006)

Kothari LS, Shah PN, Mhatre MC, "Effect of continuous light on the incidence of 9,10-dimethyl-1,2-benzanthracene induced mammary tumors in female Holtzman rats" *Cancer Lett* **16**: 313–7 (1982).

Leston J, Harthé C, Brun J, Mottolese C, Mertens P, Sindou M, Claustrat B, "Melatonin is released in the third ventricle in humans. A study in movement disorders", *Neurosci Lett* **469**:294–7 (2010).

Lin L, Huang QX, Yang SS, Chu J, Wang JZ, Tian Q, "Melatonin in Alzheimer's disease", *Int J Mol Sci* **14**: 14575–93 (2013).

Mao L, Dauchy RT, Blask DE, Slakey LM, Xiang S, Yuan L, Dauchy EM, Shan B, Brainard GC, Hanifin JP, Frasch T, Duplessis TT, Hill SM, "Photoperiodic effects on tumor development and immune function", *Mol Endocrinol* **26**:1808–20 (2012).

Nelson RJ, Blom JM, 'Photoperiodic effects on tumor development and immune function", *Bio Rhythms* **9**: 233–49 (1994).

Polimeni G, Esposito E, Bevelacqua V, Guarneri C, Cuzzocrea S, "Role of melatonin supplementation in neurodegenerative disorders", *Front Biosci (Landmark Ed)* **19**: 429–46 (2014).

Sahin L, Figueiro MG, "Alerting effects of short-wavelength (blue) and long-wavelength (red) lights in the afternoon", *Physiol Behav* **116–117**: 1–7 (2013).

Santhi N, Thorne HC, van der Veen DR, Johnsen S, Mills SL, Hommes V, Schlangen LJ, Archer SN, Dijk DJ, "The spectral composition of evening light and individual differences in the suppression of melatonin and delay of sleep in humans", *J Pineal Res* **53**: 47–59 (2012).

Schernhammer ES, Laden F, Speizer FE, Willett WC, Hunter DJ, Kawachi I, Colditz GA, "Rotating night shifts and risk of breast cancer in women participating in the nurses' health study", *J Natl Cancer Inst* **93**: 1563–8 (2001).

Schernhammer ES, Hankinson SE, "Urinary melatonin levels and postmenopausal breast cancer risk in the Nurses' Health Study cohort", *Cancer Epidemiol Biomarkers Prev* **18**: 74–9 (2009).

Stanberry LR, Das Gupta TK, Beattie CW, "Photoperiodic control of melanoma growth in hamsters: influence of pinealectomy and melatonin", *Endocrinology* **113**: 469–75 (1983).

Stevens RG, Brainard GC, Blask DE, Lockley SW, Motta ME, "Adverse health effects of nighttime lighting: comments on American Medical Association policy statement", *Am J Prev Med* **45**: 343–6 (2013).

Stevens RG, "Lighting during the day and night: possible impact on risk of breast cancer", *Neuro Endocrinol Lett* **23**: 57–60 (2002).

Thapan K, Arendt J, Skene D, "An action spectrum for melatonin suppression: evidence for a novel non-rod, non-cone photoreceptor system in humans" *J Physiol* **585**: 261–267 (2001).

CHAPTER 6

Signs of the Times

When you are old, you want things to happen quickly. Ten years ago, when we established the business selling products to protect people from the damaging effects of light at night, we expected rapid growth. Maybe because we are physicists rather than businessmen, we didn't really know how to let people know about the problem we wanted to solve. We were concerned that if we grew too rapidly, the big companies would simply take over and cut us out of the business. So, we have been content with slow but steady growth—but there are signs the public is beginning to catch on.

It was a good sign when Philips introduced the HUE lamp. It is an LED source whose color can be adjusted with a smartphone. It can become a low-blue light when only the red LEDs are lighted. Another sign was the introduction of the Goodnight bulb by Lighting Sciences, and, most recently, the ALIGN light bulbs from GE, which they offer in AM (high in blue) and PM (low in blue) models. I think making the "wake-up" bulbs a very blue white, though, is a mistake. People's experience with very blue fluorescent bulbs at their introduction made them unacceptable. On the other hand, what do I know? The Japanese apparently like the very blue bulbs.

Lighting companies' moves to introduce lights with reduced blue light is partly meant to reduce risk on their part. It is an admission that ordinary light bulbs have something wrong with them: they suppress melatonin, and so the companies risk the possibility of lawsuits from shift workers who contract breast cancer.

The following story describes something that happened in Denmark:

Denmark Pays Compensation for Breast Cancer after Night-Shift Work
Zosia Chustecka

March 23, 2009

The Danish National Board of Industrial Injuries explains on its Web site that in 2008, breast cancer after night-shift work was recognized as an industrial injury in 38 of 75 cases that were submitted to the Occupational Disease Committee. Compensation was granted in all but 1 of the 38 cases, and was paid by the employer's industrial-injuries insurance.

End of Danish news story.

We were happy when the *New York Times* ran an article in the Tuesday Science Section "Can orange glasses help you sleep better?" by Kate Galbraith. It contained a link to our website that resulted in a peak in sales.

Shortly after the *NY Times* story appeared, George Stephanopoulos interviewed Dr. Richard Besser (ABC's medical expert) on his morning show about the benefit of wearing orange glasses in the hours before bedtime. CBS also did a very brief story about orange glasses and sleep.

It is encouraging that the United States Department of Health and Human Services funded both a phase 1 and phase 2 SBIR (small business innovative research) to develop a lighting system that will not depress the production of melatonin. At least one part of the federal government recognizes the Silent Killer. The goal is to develop a ceiling lighting system in which lighted tiles will replace the ones in a drop ceiling. The principal investigator is the founder and CEO of Circadian,

Inc. Dr. Martin Moore-Ede. He has excellent credentials and taught for many years at the Harvard Medical School. The company advises companies on how to schedule shift work.

In 2013 a group of Japanese scientists came together and founded the Blue Light Society. In 2014 they sponsored the first Blue Light Symposium in Tokyo, Japan that was attended by about 300 scientists from all over the world. In 2015 they are sponsoring the 2nd Blue Light Symposium in New York City. Dr. Blask who has done such excellent work in establishing the link between light at night and breast cancer will be a featured speaker. Dr. Stevens who proposed the "Melatonin Hypothesis" to explain the rise in breast cancer, is also featured.

Regarding prostate cancer, a newly reported study showed that men whose overnight melatonin was measured some years ago were followed for many years and prostate cancer deaths recorded. Four times as many deaths were reported for those whose melatonin production was less than the mean than those whose melatonin production was greater than the mean. Since there were only 19 deaths from prostate cancer in the group, one cannot draw any firm conclusions, but it strongly suggests that maximizing one's overnight production of melatonin is worth doing. How hard is it to change to a low-blue light bulb?

CHAPTER 7

What the Reader Can Do about All This

Lists are very popular these days. Here is my list of the ways you can avoid becoming a victim of the silent killer.

Let's start with what you can do first thing in the morning. (If you are a shift worker, you can skip this part and go to the next section.)

1. If your bathroom isn't already brightly lit, buy a fixture that makes the room bright. Do the same for your kitchen or your breakfast nook. There are special lights for treating seasonal affective disorder (SAD) that can work for this. (For example, the Center for Environmental Therapeutics has one for $160.) I suggest you spend a similar amount to get a really bright fixture for your kitchen. It is a lifetime investment with a big payback. Getting bright light at about the same time every morning is the first step in developing a robust circadian rhythm. Never look directluy at the light, it can damage your eyes. That's why an overhead fixture that makes the whole room bright is the best bet.

2. The second step is to get outside as much as possible during the day. Park as far away from your office as reasonable and enjoy the sunlight. Even on the gloomiest day, the outside light intensity is higher than that of indoor lighting. It's good for your spirits as well as your body.

3. Eat healthy foods at about the same times throughout the day. Studies show that the time you eat has an impact on your circadian rhythm. It's not as strong as the timing of light exposure but not negligible.

4. Approximately twelve hours after your initial exposure to light, put on glasses that block blue light or enter an area free of blue light. This allows your body to begin the production of melatonin. Since it can flow for only about twelve hours, it will finish about the time you awaken. Without melatonin present anymore, you will wake up feeling refreshed without the need for an alarm clock.

5. A short time before bedtime, have a light snack. It is hard to sleep well if you are hungry or thirsty.

6. Keeping well hydrated is important to good health and is the easiest and cheapest thing we can do. Remember that our ancient ancestors lived in the sea. Now we bring the sea with us in the fluids in our bodies. We should always have a small surplus supply of water. When mineral concentration gets too high, solids begin to precipitate, and problems in joints (gout, bone growths or spurs) and kidneys (stones) can result. Have a small drink every time you wash your hands. Don't make your body beg for water by making you feel thirsty.

7. Go to bed at about the same time every night, and sleep in darkness. Use a sleep mask if you live where street lights or passing cars illuminate your bedroom. Use low-blue night-lights throughout the area to avoid falls.

8. Don't worry; none of these rules are engraved in stone or brought down from the mountaintop. Remember that the

circadian rhythm is really quite robust. Missing out every now and then really won't have a big impact.

9. Be happy. You are doing what any thinking person would do to stay healthy. If, despite your efforts, you still have problems, you know you have given it your best shot.

10. Share what you have learned with your friends and neighbors. They deserve to know about how easy it is to avoid the silent killer.

Shift Work

Most of the recent studies of shift work find that individuals who have rotating shifts tend not to adapt to a nocturnal rhythm. In other words, while they work nights during the week, their circadian rhythms stay the same as on the weekend, when they keep normal hours. This strongly suggests that shift workers should try to keep their melatonin flowing during the same hours as on the weekend, even while they're working. This may be made possible by wearing amber glasses while working at night.

A study in Japan suggests that preserving a shift worker's normal schedule of melatonin production might be easier than I had originally thought. The subjects in the study wore caps with a transparent red visor while working a night shift and found that their melatonin was only reduced by 8 percent compared to those who wore caps with no visor, who experienced a 52 percent reduction.

Wearing amber glasses may do a better job to preserve melatonin production, but it may attract more unwanted attention than wearing a cap with an opaque visor. If red works, opaque is at least as good. Not too long ago, nurses wore caps. Perhaps an all-white version of the caps with visors that women navy officers wear would be acceptable.

That wearing a visor is that successful in preventing melatonin suppression really should not be surprising. We have known for many years that the sensors that control the circadian system are located primarily in the lower half of the retina. Since most lighting projects from the direction of the ceiling, it makes sense that a visor can work quite well to preserve the melatonin of night-shift workers.

Amber glasses that block blue light may be of value to nurses working night or rotating shifts.

Permanent night-shift workers should try to turn their days into nights with blackout shades on their windows and should put on amber glasses for two or three hours before going to bed in darkness during the day. They should keep the same schedule on days off and weekends. This should maximize their production of melatonin and thus avoid any increased risk of breast and prostate cancer.

Shift workers on slowly rotating shifts (e.g., monthly) should try to adapt as the shifts rotate and try to use the glasses two or three hours ahead of bedtime regardless of which shift they are on. Again, they should sleep in darkness.

For quick-rotation shift workers (a week or less), it may be possible to keep the body on the same schedule as when working days. That is, put on the glasses at the same time every evening (e.g., 8:00 p.m.), whether planning to go to bed at 11:00 p.m. or to work that night. Wear them until bedtime on days off or all night while working (e.g., until 7:00 a.m.). Since melatonin is present while working nights, sleepiness is potentially a problem. However, one brief study found no decline in performance under these conditions (wearing amber glasses while working at night).

Final Thoughts

In a free society, we can do what we want. Our 24/7 lifestyle was developed in ignorance of the silent killer. While many nighttime jobs are vital, like those of nurses and firefighters, most night work is done to reduce costs for a company or for the worker to make more money. Perhaps as we grow smarter about the actual cost in human health, we will move toward minimizing exposure to light at night.